AF498401

Official Guide to the Lewis and Clark Centennial Exposition, Portland, Oregon, June 1 to October 15, 1905 ..

THE Lewis and Clark Centennial Exposition celebrates the one-hundredth anniversary of the exploration of the Oregon Country by an expedition under command of Captains Meriwether Lewis and William Clark, of the United States army. Thomas Jefferson, President of the United States, sent to Congress in 1803 a message recommending this expedition, his object being to explore the then unknown region lying between the headwaters of the Missouri River and the Pacific Ocean, as well as the Louisiana territory intervening. The President's message resulted in the appropriation of a small sum for the expenses of the expedition, and Jefferson appointed his private secretary, Captain Lewis, to command the party. Lewis chose his friend, Captain Clark, to accompany him as joint commander. Lewis and Clark outfitted for their long journey at St. Louis, gathering about thirty trusted men for the expedition. The start was made from Camp Dubois, Illinois, a few miles above St. Louis, May 14, 1804. The party ascended the Missouri River and spent the following winter amongst the Mandan Indians in what is now North Dakota. On April 7, 1805, the expedition resumed the voyage up the Missouri to a point near the headwaters of that stream, and traveled by land through the mountains. They went down the Columbia River to its mouth, reaching the Pacific Ocean November 7th, having traveled more than 4,000 miles. Spending the winter in a log stockade which they built on the south side of the Columbia, in what is now Clatsop County, Oregon, they started upon the return journey March 23, 1806, reaching St. Louis September 23d, after an absence of two years and four months. They were the first Americans who crossed the continent.

Historical Significance

The success of the Lewis and Clark expedition enabled the United States to acquire all the territory now embraced in Oregon, Washington and Idaho and the western parts of Montana and Wyoming.

Meriwether Lewis
Wm Clark

Plan and Scope of the Exposition

THE Lewis and Clark Centennial Exposition and Oriental Fair, the first international exposition under the patronage of the United States Government ever held west of the Rocky Mountains, is a demonstration of the marvelous progress of Western America and an exemplification of the great possibilities for trade development in the Orient. The exposition is a world's fair in every sense, all of the chief nations of Europe and the Orient being participants, and many of the American states taking part. The Western states, from the Rocky Mountains to the Pacific, are participants upon a larger scale than has been the case at any previous exposition. Each of the Pacific Coast states is represented by exhibits which, for the first time, show to the world comprehensively the marvelous achievements of the Coast country, its wonderful natural riches and the vast wealth that still lies undeveloped in these states. The exposition is built upon the idea of compactness without crowding, and with a view to the comfort and convenience of the visitor, rather than to present an aspect of mere bigness. The grounds, comprising 406 acres of land and water, are large enough to accommodate the choicest of exhibits from all over the world, and small enough to enable the visitor to study the entire exposition without becoming footsore and weary. It is the belief of the men who made this exposition, seconded by many expert "expositionists," that in the laying out of the grounds and the placing of the various exhibit and amusement features the nearest approach to perfection ever achieved by a similar enterprise has been made. This is an exposition of processes as well as of products. In addition to the raw materials, the processes by which they are converted into things of worth and beauty are shown and the finished products are also seen here. The Lewis and Clark Centennial Exposition is, in short, a world-epitome of the arts, industries and general achievements of mankind.

GRAND STAIRWAY
SACOJAWEA.
PERISTYLE GOVERMENT BLDG.
SUNKEN GARDENS.
COLONNADE ENTRANCE AND ADMINISTRATION BLDG.

United States Government

THE United States Government Buildings are situated on the peninsula projecting into Guild's Lake and may be reached by the Grand Esplanade, the Trail and the Bridge of Nations, or by one of the numerous boats which convey visitors to the landing facing the main government structure. A large part of the $450,000 appropriation has been utilized for the finest buildings ever constructed for government participation in an exposition. In design the structures are of Spanish renaissance. Two towers, 260 feet high, surmount the main building. In this are housed the displays of the various departments of the government. The Irrigation building and Territorial structure are connected with the main building by classic peristyles. These contain comprehensive exhibits of the latest irrigation projects undertaken by the government in the great arid sections of the country, and exhibits of the products and manufactures of our territories. The Fisheries exhibit is located in a building erected for the purpose in the rear of the main building. In this are shown the various methods of maturing different species of fish that frequent the streams of the Pacific Northwest. The vertebrates are stripped of their eggs. These are jarred and processed. All the stages of the growth of the fish are shown in view of the public. The Life-Saving Station is located to the left of the main group and on the shore of the lake. Exhibition drills are given daily. To the right of the main building are the camps of the military, marine and signal corps, as well as the Heavy Ordinance exhibit, in which are shown the large guns utilized for coast defenses and siege operations. A feature of this exhibit is a full-sized 10-inch sea-coast gun, mounted behind ramparts on the latest pattern of disappearing carriage. On the outside are the many totem poles and quaint boats, too cumbersome to put inside the building. A life-saving station, where daily drills are given, is also found on the island.

UNITED STATES GOVERNMENT BUILDING

State Buildings

PARTICIPATION by Oregon's sister commonwealths has been exceedingly liberal. A number of splendid state buildings enhance the general appearance of the exposition. California has erected a picturesque structure just in the rear of the Forestry building. It is a replica of four of the famous Franciscan missions. Adjoining this is the Washington building, a magnificent example of classical architecture and one of the largest state buildings on the grounds. Facing the Manufactures, Liberal Arts and Varied Industries palace is the Missouri state building. This is a handsome structure and contains a representative exhibit of Missouri's products and manufactures. The Oregon building is easily reached from the European Exhibits palace and the main entrance. In design the structure is of Colonial architecture. The Massachusetts building, which adjoins that of Oregon, is a reproduction of the famous Bullfinch front of the State House on Beacon Hill, Boston. New York's building occupies a commanding site on Lakeview Terrace, overlooking the lake and government peninsula. This building is designed after the Italian renaissance and is one of the handsomest of any of the state buildings. Idaho has built a unique building of a style peculiar to inter-mountain countries, immediately adjoining the New York building. Illinois' building also forms one of this group, as well as the Utah state building. Excepting Illinois, each state building contains a comprehensive display of the respective states' resources, products and manufactures. Aside from exhibit purposes the buildings have been designed to afford ample accommodations for the entertainment of visitors. The Illinois building is for entertainment purposes only. All the state buildings are conveniently located. At previous expositions the state buildings have been located in remote parts of the grounds. The total state participation aggregates more than one million dollars.

WASHINGTON BLDG.
CALIFORNIA BLDG.
NEW YORK BLDG.
MISSOURI BLDG.
OREGON BLDG.

Agriculture Building

THIS structure is the largest on the grounds and stands to the right of Columbia Court as you pass through the Colonnade entrance. Its huge translucent dome may be seen not only from every part of the exposition grounds, but from many points in Portland and for miles down the Willamette River. In design the structure is a composite study of the Spanish renaissance and Mission architecture. The building contains one of the most thorough displays of agricultural and horticultural products, foods and food products ever offered to public view, and affords those unfamiliar with agricultural and horticultural fields an excellent opportunity to obtain a more intelligent conception of these industries. Oregon has acquired a great part of the space for exhibit purposes. All displays are made by counties. In this manner comprehensive exhibits of the products peculiar to the different sections of the state are made. The marvelous apples and strawberries of the Hood River district are creditably represented. From the Willamette Valley and Southern Oregon various products characteristic of those regions are exhibited. A thorough display is made of the products of the great arid tracts in Eastern Oregon that are under irrigation. California and Washington have large exhibit spaces which accommodate the displays that could not be contained in their state buildings. Manufacturers throughout the country make attractitve displays of manufactured food products, and in some instances show the practical side of their processes with intricate mechanisms in operation. Fine wines form an important part of the exhibits of the Agriculture building. Wine-makers of California, famous for the excellence of their products, are well represented. Eastern as well as Western beer manufacturers make large displays. Whiskey manufacturing is illustrated with interesting exhibits of the processing, the crude and finished products. Some of the foremost distillers in the country are participants.

AGRICULTURAL PALACE

Statuary

MANY groups of paramount historical interest are included among the sculptural offerings of the exposition. The compositions are all by American sculptors and present one of the finest exhibits ever included in a like enterprise. Just inside the imposing Colonnade entrance is a group by Frederick Remington, who has portrayed the playful side of the life of the cowboy. Four hilarious "cow-punchers," possibly a little the worse from an over-indulgence in liquor, are depicted on the backs of their favorite bronchos, dashing furiously through a town and shooting their revolvers in the air. This group is typical of the earlier Western life, when Saturday night's diversion was to get drunk and "shoot up the town." Standing in the center of Columbia court is a heroic bronze figure of Sacajawea, the Shoshone woman who guided Lewis and Clark on their famous expedition into the great Northwest. The composition is by Miss Alice Cooper, of Denver, and shows a conscientious study of the theme. Sacajawea is portrayed as a young mother, her papoose lashed to her back. With one arm raised high and face radiant, it would seem that after months of awful suffering she found solace in her first glance at the placid Pacific. The figures of Captains Meriwether Lewis and William Clark, which adorn either end of the balustrade on Lakeview terrace, are appropriate companion statues to that of Sacajawea. F. W. Ruckstuhl, in his composition of Captain Clark, portrays the explorer as he was garbed upon the completion of his long journey. The figure of Meriwether Lewis is by Charles Lopez. Several groups by Solon H. Borglum, idealizing the life of the Indians, are placed to advantage on Lakeview terrace. "A Step Toward Civilization," by A. A. Weinmann, the youngest American sculptor, is located in the Concourse plaza, and is a marvelous sculptural conception. Mr. Weinmann offers a careful study of the poetic side of his subject. The "Horse Facing the Storm" is just west of the Grand Stairway.

COWBOY AT REST.
HITTING THE TRAIL.
A PIONEER.
COW KILLED BY LYNX.
BULL FIGHTING LYNX.

The European Exhibits Building

ACROSS Columbia Court and opposite the Agriculture building is the European Exhibits building. While not so large as the neighboring structure, the building is equally as attractive, if not more so, with its numerous towers and the great central pavilion which surmounts it. The building is in general keeping with the other structures, being of Spanish renaissance in design. One-half of this entire structure is given to Italy. This country has utilized most of its space for a display of fine marble statuary. This exhibit is valued at more than a million dollars and is the best Italy has ever made. Among the many sculptural offerings are compositions of Italy's greatest masters, gathered from the Loggia in Florence and from the gallaries of Rome, Venice, Naples and Genoa. The rest of the Italian participation is made up of displays from various sections of the country, showing its manufactures and its agricultural and horticultural products. England is represented with a very creditable and interesting exhibit. France makes a splendid showing of fine arts, sculpture, costly gowns and lingerie. Germany has sent a very comprehensive exhibit, which is contained in this building. Austria's display is composed principally of a valuable exhibit of fine Bohemian glassware. Norway and Sweden have a joint display. Holland's section is devoted to an exhibit of products and manufactures peculiar to that country. The display from Prussia is complete and of unusual interest. Switzerland makes a very creditable exhibit of fine watches and fancy laces. Russia, after reconsideration, sent a splendid exhibit of fine furs and unique hammered brass and copper goods, as well as a thorough display of its agricultural products. For the convenience of visitors, all European exhibits have been housed in one building, which is amply large to accommodate them all. This greatly facilitates the locating of any exhibit and is a feature not heretofore included in an exposition.

EUROPEAN EXHIBIT PALACE

The Oriental Exhibits Building

BETWEEN the European Exhibits building and the Forestry structure, and facing the Oregon and Massachusetts buildings, will be found the Oriental Exhibits building. The main entrance to this structure is situated on Lewis and Clark boulevard. This building in design is the nearest approach to classical architecture of any of the exposition buildings. It is a free study of the Corinthian order. The main entrance is very imposing, a great arch located between massive buttresses. Ornate staff work has been used to advantage to elaborate the entire building. The long facades are broken occassionally by fluted pilasters, and all windows are surrounded with attractive casings. Japan occupies the largest space in the building, making a special display of Japanese arts and crafts. Fine silk embroidered work is featured, and bronze of the most marvelous workmanship are given advantageous places. An interesting display is made of Japanese potteries, as well as the various crockeries of the Island Kingdom. The exhibit of cloissone, democene, arita and Imari ware is of unusual interest. China follows with an exceedingly large exhibit. In this, the remarkable inlaid and hand-carved ebony furniture for which the country is famous is featured. Teakwood boxes and heavy chests are also shown, as well as a very thorough display of fine embroideries and art silks. India's display is composed principally of the wonderfully woven rugs and shawls. Aside from these, however, curiously carved ivories and antique bronzes are shown. Ceylon makes a feature of her teas, which have gained her worldly renown. Algeria and Turkey have made a joint exhibit, which was assembled under the direction of one commissioner. The display represents principally the fine sandalwood carvings and trinkets for which these countries are noted. Like the European exhibits, all Oriental displays are housed in one building for the convenience of visitors.

ORIENTAL EXHIBIT PALACE

Manufactures Liberal Arts and Varied Industries

THE building in which are housed the exhibits classified under these heads is one of the largest of the exposition palaces. It faces the Concourse Plaza and is neighbor to the Agriculture building. Spanish renaissance has been followed in the design of this structure. An imposing entrance, elaborated with Corinthian columns and pilasters, features the west facade. To facilitate the locating of any one exhibit, the displays are all grouped under separate heads. In the Graphic Arts section, a comprehensive display is made of everything pertaining to modern printery. Large press and paper manufacturers have participated liberally. The manufacture of ropes and hawsers is shown in the Cordage section. The Pharmacy section contains exhibits of the manufacture of patent nostrums, anesthetizing and pharmapeutic apparatus and medical instruments, as well as showing displays of medical supplies. This exhibit should be of unusual interest to men in the profession. The latest improved methods of systematizing offices are shown in the Office Supply section. Great files and other office furniture are shown, typewriters, adding machines and all the kindred articles that serve to make the furnishing of an office up-to-date. The exhibits showing the manufacture of cloths, threads and commodities used in tailoring and haberdashery are all grouped conveniently on the east side of the building. A number of manufacturers have erected large machines, showing the various stages of manufacture. A splendid display of American manufactured musical instruments is shown in the Musical section. An enterprising stove manufacturer offers a large and instructive exhibit showing the manner in which the family range is made. A large space is devoted to fine furniture and its manufacture. Jewelry is well represented. The exhibits in this building represent the entire participation in the Centennial by American manufacturers.

MANUFACTURES, LIBERAL ARTS AND VARIED INDUSTRIES BUILDING

Forestry Building

MOST original of all the exposition buildings is the Forestry structure. It is distinctly a Western building and represents one of Oregon's greatest industries. Constructed entirely of huge logs, it is exceedingly picturesque, with its crude facades and roughly hewn embellishments. Miles of logs, more than six feet in diameter, and tons of shakes and shingles were required for the construction of this unique structure. As to its interior, the Forestry building is inspiring. More than fifty great tree columns, veterans of a virgin forest, support the rafters. These range from six feet to eight feet in diameter and are fully sixty feet tall. The interior of the building is as crudely finished as the exterior. Here and there rustic stairways lead to the second and third floors. From the balconies great doors open upon attractive porticoes. The building is the most original architectural creation ever offered to public view and is in many respects the most interesting structure on the expostiion grounds. In this building are housed various displays showing the manner in which commercial lumber is manufactured. In one corner is a saw-mill in operation. This shows the great trees as they are fed to the saws and squared. The cutting of rough lumber is next shown, and so on until the complete process is illustrated. In another section of the building the manufacture of finished lumber and veneers is shown. Lumber products, such as doors, sashes and wainscots are exhibited. In the Oregon section are shown displays of every kind of lumber that may be found in the state. As fisheries are closely associated with the forestry industry, the state maintains a salmon hatchery in the building. This affords the visitor an opportunity to witness the cultural work among the Royal Chinook and steel-head salmon. All in all, the Forestry exhibit is the most complete of its kind ever contained in an exposition. It is easily recognized by its rustic structure.

FORESTRY BUILDING

ALMOST adjoining the Mines and Metallurgy building will be found the Machinery, Electricity and Transportation building. The construction of this building was authorized about the time the other exposition buildings were completed, and little attempt has been made at design. Notwithstanding, the structure is very presentable. The exhibits are divided into three sections. The Transportation section is in the north end. Here is shown every mode of conveyance. Automobiles are given liberal space. This display shows all the numerous types of automobiles. Gasoline propelled vehicles occupy most of the space, but electric and steam automobiles are displayed. Various types of boats are exhibited. Kindred manufactures, such as harness, saddlery and vehicle furnishings, are shown. Railroad transportation is creditably represented. The largest locomotive ever constructed in the United States forms a part of the display. Offering an interest and abrupt contrast to this gigantic Mogul is the "Oregon Pony," the first locomotive used in this state. This diminutive railroad engine is only thirteen feet long. Various types of passenger and freight locomotives are displayed, as well as refrigerator cars. Mining locomotives are also shown. The latest improved patterns of electrical appliances are shown in the Electricity section. Huge dynamos and motors are operated. Electricity in its various applications is illustrated. The telephone and telegraph and wireless telegraphy are exhibited. Hundreds of kinds of electric lights, heaters and other household commodities are displayed. In the Machinery section may be seen great engines in operation; not only steam engines, but those run by gas and compressed air. Aside from these displays, a comprehensive exhibit is made of varied transmissions, gears, boiler-making and fittings. The display of agricultural machinery is contained in an annex erected in the rear of the Transportation section.

Machinery Electricity and Transportation

MACHINERY, ELECTRICITY AND TRANSPORTATION BUILDING

Mines and Metallurgy

THE display of minerals, precious stones and mining machinery at the exposition surpasses similar exhibits at earlier expositions. The Mines and Metallurgy building is a splendid structure, 200 feet long and 100 feet wide, with a one-story addition, known as the Black Sands annex, extending 60 feet from the east facade. As the exposition is being held with the principal object of exploiting the resources of the Northwest, it is fitting that a thorough display of the mineral wealth be made. Housed in this building are exhibits from the states of Montana and Wyoming, as well as those of Oregon and exhibits from numerous individual miners illustrating the various methods of obtaining minerals from the ore. The displays of Washington, Utah and California are housed in the respective state buildings. The center of the building is a great display of gems of fabulous value, which should prove of unusual interest to the visitor. These precious stones are shown in two immense safes, fitted with plate glass fronts so that the visitor to the Mines and Metallurgy building may have an unobstructed view of the display. In one corner of the structure a large ore furnace in operation is shown. Placer as well as quartz mining is given considerable space, and the enormous monitors playing their great and powerful streams of water upon improvised mountain-sides offer an enlightening idea of the manner in which minerals are obtained by placer methods. The Black Sands annex is an entirely new exposition feature. The United States government has provided for careful and systematic investigation of these sands, which are peculiar to the Pacific Coast, to establish credence to the reports of geologists and mineralogists which would indicate that these sands possess marvelous wealth in new and valuable minerals. The building for the carrying on of the experiments represents the exposition's co-operation with the government.

MINES AND METALLURGY BUILDING

Fine Arts

THE display of fine arts is located in a fire-proof structure facing the Forestry building. This building is planned in the form of an "L" and is designed to display the numerous works of art to their best advantage. Light is afforded by a system of electric globe installations placed about seven feet from the floors of the galleries. Notable among the works displayed are examples of the early Dutch, French and English masters. These constitute groups of representative and carefully selected works. One of the most excellent features of the exhibit is a large group representing the works of Mauet, Monet, Sisley, Pissaro, Reouard, Huguet, Boudin and Mary Cossat. Another of the best groups is that of the works of George Innes. The chief feature of the Fine Arts display is its compactness. Visitors to former expositions were awed by the extent of the displays, and in many instances were forced to leave without seeing the entire exhibit. With the Fine Arts display of the Lewis and Clark Exposition what is lacking in quantity is more than made up in quality. Of the works of the latter-day painters and sculptors, a great many of well-known American geniuses are represented with creditable displays, as well as a number of the foremost men of European countries. A display which, though not housed in the Fine Arts building, may appropriately be considered an adjunct to the main exhibit, will be found in the European Exhibits building in the Italian section. This offers some of the most marvelous achievements of marble sculpture known in the field of art. The various groups and figures have been carefully selected from the galleries of Rome, Naples, Florence and Venice, and brought to the exposition as a part of the Italian government's exhibit. No more complete or creditable exhibit of fine arts has ever been made. The displays will be found classified according to countries from which they have been procured, and representatives of each will give the people courteous attention.

Livestock

THE livestock sheds are located on the peninsula projecting into Guild's lake, near the Government buildings. This display is one of the best of its kind ever made and should prove of unusual interest, particularly to stock enthusiasts. More than $50,000 has been appropriated for premiums by the Exposition Company and by various stock associations. The American Jersey Cattle Club appropriated $1,000, the American Shorthorn Breeders' Association $3,000, the Percheron Registry Association $2,000, the American Hereford Breeders' Association $1,000, and numerous other organizations have made allotments of smaller denomination. The Livestock display is by far the largest ever held on the Pacific Coast, and will afford the enthusiasts of the Pacific Northwest an opportunity to see the best cattle from every part of the country. All cattle participating are registered. A representative display of Holsteins is made. Record-breaking milk and butter makers, which have astonished the entire dairying world, are to be seen in the sheds on the peninsula. Jerseys participate to a large extent, as well as Herefords and Durhams. The dates for the cattle show are from September 9th to September 29th, inclusive. The show of sheep, goats and swine will also be made within these dates. From Oregon in particular an excellent showing of hogs is made. The horse exhibit will be from August 28th to September 8th. The poultry show is from October 5th to October 12th. In this, egg-laying prodigies of every known specie will be in evidence, and the finest of game cocks will be displayed. The location for the Livestock display is easily accessible from various parts of the grounds —from the Government buildings by a short walk, from the Trail over the Bridge of Nations, or from the Grand esplanade by means of the numerous boats on Guild's lake.

Public Convenience Arrangements

IN pursuance of its policy of providing every possible comfort to the visitors at the exposition, the Lewis and Clark fair management has made unexcelled public comfort arrangements by building commodious toilet rooms in every exhibit palace, and other parts of grounds. There are in all eighteen of these rooms, all handsomely and conveniently appointed.

Contrary to the custom of previous expositions, there is no charge for public convenience, except that a fee of five cents is charged for use of towels, mirror, soap and brushes. At each place there is also a shoe-shining stand for the convenience of visitors at the fair.

The buildings in which toilets are found are as follows: Forestry, Oriental Exhibits, European Exhibits, Agriculture, Liberal Arts, Mines and Metallurgy, and Electricity and Transportation. Public convenience arrangements are provided also, both for men and women, in the Experimental Gardens. In addition to these provisions, every state building is similarly equipped; and a majority of the concessions on The Trail have toilet rooms. In all cases the service is free.

Portland has one of the best water systems in the world, and an abundance of clear and sparkling water is provided free at all times at many places on the exposition grounds. Portland water is piped from Bull Run, a small stream which has its origin in the glaciers on Mt. Hood, and its absolute purity has been established by expert analysis.

In Centennial park, a charming piece of woodland in the western portion of the grounds, visitors may rest on rustic benches which line the avenues, while in various other parts of the grounds, especially on the slope of Lakeview terrace, comfortable benches have been placed at frequent intervals. The terrace benches are especially popular, as band concerts are given every afternoon and evening in the music shell on the shore of Guild's lake.

THERE are few, if any places, in the State of Oregon where as many opportunities are offered the homeseeker and investor as in Jackson County, of which Medford is the principal city.

The Medford people will be found ready to aid visitors in seeing their resources. A handsome exhibit building has been erected near the depot, and people passing through can get off and see her farm products exhibited and form their own conclusions. This section of the country, while it cannot be excelled as a grain producing country, has been found even more valuable as a fruit and dairy country. The fruits raised here are world famous, and the rich bottom lands will produce three and often four large crops of alfalfa without irrigation.

Under the able management of Medford's leading citizens a railroad has been projected from Medford to Crater Lake. The construction has already been begun, and grading is well under way towards Eagle Point, some 12 miles from Medford. From Eagle Point the road will be pushed as rapidly as possible until the timber belt is tapped, after which extensions will be made as fast as they are justified.

It is not so much the opening up of these vast pine forests, with their 4,000,000,000 feet of timber, however, which is of the greatest importance to the valley, but the fact that in addition to furnishing transportation facilities to the lumbermen, the road will enable the thousands of acres of unimproved agricultural land to be utilized for farming and fruit raising, and will also, by allowing the timber to be cut, convert 250,000 acres of timber land into the best fruit land, for the soil in the forests of the valley and foot hills, unlike that of many portions of the State, is very rich, and will produce all the varieties of vegetation which the rich bottom lands produce, and at the same time has the advantage of being naturally provided with water in abundance for irrigation purposes, which, while not necessary, enables the farmer and fruit grower to raise much larger and more valuable crops than can be raised without irrigation, although the natural production, without irrigation, far surpasses that of any other part of the State.

The importance of this feature of the enterprise cannot be overestimated, as the thousands of homeseekers from the East who are looking for locations in the West, in order to escape the severe Winters and scorching Summers of the Middle States, with their droughts and crop failures, are in most cases people of moderate means, who cannot afford to invest in the high priced improved lands of the rich valleys of Oregon, and yet do not desire to locate beyond the reach of railroad transportation. The opening up of these great forests to the farmer and small rancher by cutting off the timber, will meet both of these objections, as the owner of the land will already have made a profit on the land out of the timber, and so be enabled to sell the bare land at low prices, but the newcomer will have railroad facilities almost at his door, and be enabled to dispose of his products in the best markets. Another advantage of special interest to the fruit grower is the fact that the orchards planted in that section will be entirely free from fruit pests of all kinds for years to come, while the land is even better adapted to fruit culture than the lands in the center of the valley.

Columbia River U. S. Government Exhibit Mt. St. Helens

Trail and Bridge of Nations

American Inn

Experimental Gardens

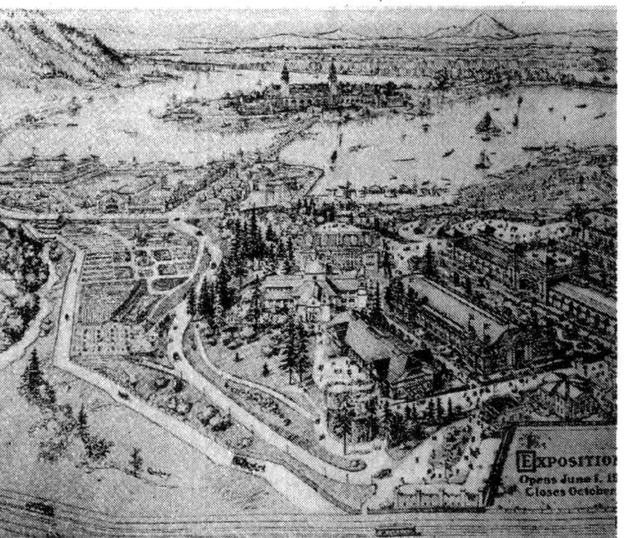

Forestry Building

Fine Arts Building

Europea

Orienta

Willamette River Mt. Hood

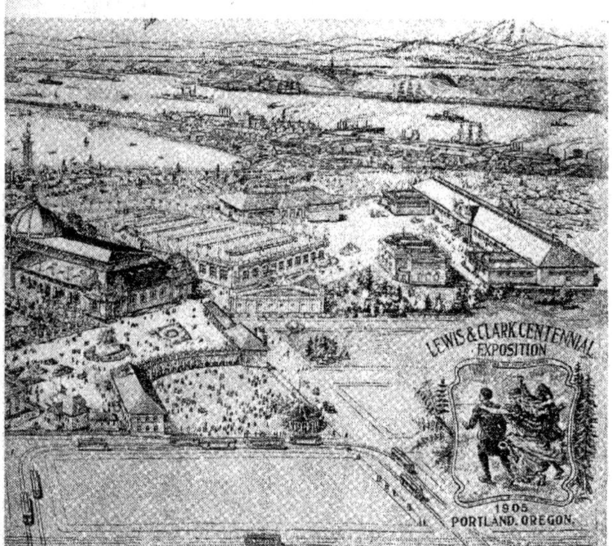

Mining Building

Machinery Building

Varied Industries Building

Agricultural Building

ace Main Entrance

ce

AUDITORIUM OR FESTIVAL HALL

Joy of a Lifetime

Take Steamers
Bailey Gatzert
Dalles City
Regulator

Trip up the Columbia River

Leave Alder Street Dock every day.

Phone Main 914

S M. McDonald
Agent

THE PRINCIPAL EXHIBITORS AND WHERE TO FIND THEM

Manufactures, Liberal Arts and Varied Industries Building

Acme Tiiturator, Los Angeles, Triturators
Allen & Gilbert-Ramaker Co, Portland, Musical Instruments
American Art Leather Co, Los Angeles, Leather Goods, Concession.
American Electrical Novelty Manufacturing Co, New York, Electrical Novelties Concession
American Numismatic Association, St. Louis, Concession.
American Sales Book Co, New York, per Wilcox, Portland, Books
American Woolen Mills Co, Providence, Woolen Goods.
Archer & Schanz, Portland, Chewing Gum
Arithmo Adding Machine Co, Detroit, Adding Machines
Arithmo Game Boards, St Louis, Game Boards
Art Crafts Shop, Buffalo, Art Furniture
Banister Co., J A., Newark, N. J., Shoes
Bannockburn Mfg. Co, Portland, Loom
Barchus, Eliza R, Portland, Pictures Concession
Bayer, J. C, Portland, Furnaces.
Bell, Geo Co, Denver, Lapidary
Bill, Edw L, New York, Musical Publishers
Blumauer-Frank Co, Portland, Chemicals
Born Steel Range Co, per M Seller & Co, Portland, Ranges
Bowers Rubber Co, San Francisco, Rubber Goods.
Broderick & Bascom Rope Co, St. Louis, Cables, Pulleys, etc
Bushong & Co, Portland, Printing in Operation
Buster Brown Stocking Co., New York, Stocking Mfg.
Cable Co, Musical Instruments.
Chamberlain Metal Strip, Detroit, Window Strips.
Charter Oak Co, per Hexter, May & Co, Portland, Ranges.
Church Co., John, Musical Instruments
Claus Shear Co, Fremont, O, Cutlery Concession.
Columbia Phonograph Co, Portland, Phonographs
Columbia River Paper Co., Portland, Paper Pulp
Comptograph Co., Chicago, Comptograph.
Crane Bros, Westfield, Mass., Ledger Paper.
Deasey Water Heater Co., Los Angeles, Water Heaters.
Dockerty, A J, Portland, Ethnological, Indian Novelties, and Blanket Weaving.
Dodd & Mead, New York and Chicago, Publications
Doter Mfg Co, per Geo. F. Eberhard & Co., San Francisco, Sad Irons.
Edison Mfg Co, Orange, N. J., Phonographs & Moving Pictures.
Emerick Co, Chas, Chicago, Feathers.
Erickson, E. H, Artificial Limb Co, Minneapolis, Artificial Limbs.
Fairbanks, N K, Chicago, Soap.
Fry, H. C., Glass Co, Rochester, Pa., Electrical Novelties Concession
Gill, J. K, Portland, Publications and Stationery.
Gillette Safety Razor Co., Chicago, Patent Razors.
Greenwald, O., St Louis, Linen Concession.
Grossman, H, Florida, Alligator Goods Concession
Gurney, Edmund, Portland, Ladies' Dresses
Harris Trunk Co, Portland, Trunks.
Hawes Von Gal. Co, Danbury, Conn, Hat Factory.
Hepburn & Morrison, St. Louis, Chain Curtains.
Hilbert, A. J. Co, Milwaukee, Perfumery.
Hunter Arms Co, Fulton, N Y., Fire Arms.
I. & C. Fisher, Musical Instruments
Ingersoll Watch Co, Chicago, per Lipman-Wolfe, Watches.
Irwin-Hodson Co, Portland, Printing in Operation.
Ishiguro, C. I, San Francisco, Photographs.
Kampfe Bros, New York.

Continued on page 40

MANUFACTURES, LIBERAL ARTS AND VARIED INDUSTRIES BUILDING—Continued from Page 38

Kilham Stationery Co., Eastern Firm, Stationery Supplies and Publications.
Koeckeritz, F., St. Louis, American Pearls.
Laird & Schoeber, Philadelphia, Shoes.
Lambert Pharmical Co., St. Louis, Chemicals.
Lewis-Stenger Barber Supplies, Portland, Barbers' Supplies.
Libby Glass Co., Chicago, Cut Glass, per Olds, Wortman & K.
Libby, McNeil & Libby, Chicago, Music Binding.
H. Liebes Bros., Portland, Furs.
Lutke Mfg. Co., Portland, Show Cases.
Malleable Steel Range Co., per H. E. Edwards, Portland, Ranges.
Manion, J. J., St. Louis, Operating Loom Concession.
Marks, A. A., New York, Artificial Limbs.
Marks' Adjustable Chair, New York, Adjustable Chairs, per Beede, Portland, Adjustable Chairs.
McLynn Pulley & Patent Co., Portland, Models.
Meier & Frank Co., Portland, Ribbons.
Mergenthaler Linotype Co., New York and San Francisco, Linotype Machines.
Miehle Printing Press & Mfg. Co., Chicago, Printing Presses.
Miller Keyless Lock Co., Kent, O., Patent Locks.
Minnesota Desk Mfg. Co., Minneapolis, Office Furniture.
Mutual Label & Lithograph Co., Portland, Lithographic Work.
National Cash Register, Dayton, O., Cash Registers.
Norris Safe & Lock Co., Portland, Safes.
Northwest Oil & Paint Co., Portland, Paints.
Oliver Typewriting Co., Seattle, Typewriters.
Olympia Drug Co., Seattle, Chemicals.
Oregon Dental Supply Co., Portland, Dental Supplies.
Oregon Furniture Mfg. Co., Portland, Furniture.
Oregon Optical Co., Portland, Field and Opera Glasses, etc.
Optical Concession.
Pacific Coast Agency, Portland, Typewriter Supplies.
Pacific Coast Rubber Co., Portland, Rubber Goods.
Pacific Regalia Co., Portland, Fraternal Buttons, etc.
Packard Co., Musical Instruments.
Pacific Stamp Works, Portland, Rubber Stamps.
Parker Bros., Meriden, Fire Arms.
Peters Cartridge Co., Cincinnati, Ammunition.
Pfunder, Dr., Portland, Patent Medicines.
Portland Cigar Co., Portland, Cigars, Concession.
Portland Cordage Co., Portland, Rope Making Mfg.
Portland Stove Works, Portland, Ranges.
Portland Woolen Mills Co., Portland, Looms.
Premier Piano Player Co., San Francisco, Piano Player.
Pyle James, & Sons, New York, Pearline.
Rand & Reed, Worcester, Taxidermy
Rich, B. B., Portland, Curios Concession.
Silverfield Co., Portland, Furs.
Singer Sewing Machine Mfg. Co., New York, Sewing Machines in Operation.
Standard Fire Apparatus, Seattle, Apparatus.
Stetler, F. C., Portland, Paper Boxes.
St. Louis Art Leather Co., St. Louis, Art Leather.
Sweeny Surgical Co., Los Angeles, Surgical Appliances.
The Ideal Co., New York, Eye Rest.
Toledo Cooker Co., Toledo, O., Cooking Utensils.
Tower Co., A.J., Boston, Oil Clothing
Tuerk, I. K., Portland, Artistic Metal Work.
Tull & Gibbs, Portland, Furniture.
Underwood Typewriter Co., New York, Typewriters.
Van Vleck, Chas. H., Kansas City, Patent Medicines.
Victor Kremer, Chicago, Music Selling Concession.
Walden Knife Co., New York, Cutlery Concession.
Wall, Wm. E., per Sutcliff & Blied, Portland, Grained Wood.
Warren, W. R. & Co., Philadelphia, Chemicals.
Weller, S. A., Zanesville, O., Art Pottery.
Wells, J. L., Taunton, Mass., Aluminum Goods.
Western Mfg. Co., San Francisco, Models.
Wheeler & Wilson Mfg. Co., St. Louis, Sewing Machines.
Woodard, Clarke & Co., Portland, Cosmetics, Chemicals, Concession.
Yucca Artificial Limb Co., Los Angeles, Artificial Limbs.
Zan Bros., Portland, Brooms.

Continued on page 42

Hill Military Academy as a Hotel, 1905

VISITORS to the Lewis and Clark Fair at Portland will find homelike accommodations at **The Hill Military Academy.** The large permanent buildings have been specially rearranged for this purpose. The location in the best residential portion of the city and within easy walking distance of the Fair is ideal for the entertainment of Fair visitors. It is the only building in the block and commands from all sides a free outlook and excellent view. Direct car service from Union Depot. Meals a la carte. European plan. Rates, $1.00 and up. For rates and reservation apply to J. W. HILL, Proprietor, 821 Marshall Street, Portland, Ore.

WHY not send your boy to a military school? Investigate the advantages offered by **The Hill Military Academy.** Illustrated Catalogue and Booklet on application to J. W. HILL, M. D., Proprietor and Principal,

821 Marshall Street, Portland, Ore.

Transportation, Electricity and Machinery Building

Continued from Page 40

TRANSPORTATION SECTION.

Beryman Leather Co., Portland, Leather.

Campbell, W. C., Seattle, Doubletrees.

Columbus Buggy Co., Columbus, Vehicles.

Cronin, P. J., Co., Portland, Saddlery and Harness.

John Deere Plow Co., Portland, Implements.

Keats, H. L., Auto Co., Portland, Concession.

Lawrence Co., Geo. W., Portland, Saddlery and Harness.

Mitchell, Lewis & Staver, Portland, Collective.

Neff, W. B., St. Louis, Collective Exhibit.

Portland Implement Co., Portland, Implements.

Still Pneumatic Horse Collar Co., Bloomington, Patent Collars.

St. Louis, Building Material.

Studebaker Bros., N. W., Portland, Vehicles.

Waterhouse & Lester, Portland, Blacksmith Tools.

Western Wheeled Scraper, Aurora, Ill., Road Machines.

TRACK

Burnham, Williams & Co., Philadelphia, Baldwin Locomotives.

Climax Mfg. Co., Seattle, Logging Engines.

Hoffman Mfg. Co., Sandusky, O., Patent Trucks.

Kullman, Salz Co., Leather.

Lima Locomotive Co., Lima, Locomotives.

Railway & Steel Supply Co., Seattle, Logging Engines.

St. Louis Refrigerator, St. Louis, Refrigerators.

Oregon Pony.

ELECTRICITY SECTION

American School of Correspondence, Chicago, Technique Literature.

General Electric Co., Schenectady, N. Y., Electrical Machinery.

Northern Electric Co., Seattle, Electric Appliances.

Pacific Tel. & Tel. Co., Portland, Telegraph and Telephone Appliances.

Ricards, Wm., Portland, Electric Machines.

Standard Underground Cable Co., Pittsburg, Electric Coils and Cables.

Western Electric Co., San Francisco, Electrical Machinery.

Westinghouse E. & Co., Mfg., Seattle, Electrical Machinery.

MACHINERY SECTION

Aultman & Taylor, Portland, Traction Engine.

Averil Machine Co., A. H., Portland, Engines.

Borquist, Portland, Logging Tools.

Bulls Head Oil Co., San Francisco, Machine Oil.

Columbia Engineering Works, Portland, General Machinery.

Dodd, C. H., Astoria, Tin Can Co.

Dodge Mfg. Co., Mishawaka, Machinery.

Fairbanks, Morse & Co., Chicago, General Machinery.

Gauld & Cline, Portland, Plumbing Fixtures.

Graton & Knight Mfg. Co., Worcester, Leather Belting.

Meese, Gottfried Co., San Francisco, Power Transmitters.

Powell Co., Wm., Cincinnati, Valves.

Roebling, J. A., Sons Co., Trenton, Machine Wires.

Sonders & Co., San Francisco, Copper Stills.

Tatum & Bowen, Portland, Mine Machinery.

Willamette Iron & Steel, Portland, Collective.

Continued on page 44

Studebaker

Carriages, Wagons Harness, Automobiles

Largest Stock of Vehicles Carried on the Pacific Coast

Repository: 330-336 East Morrison Street :: :: :: Portland, Oregon

Studebaker Bros. Co., Northwest

Oriental Building—Concluded from Page 42

GALLERY

American Humane Association, Wyncote, Pa., Society Literature.
Christian Science Pub. Society, Boston and Portland, Religious Literature.
Dobbs, B. B., San Francisco, Photographs, Alaska Views.
Fisk Teachers' Agency, Portland, Educational Literature.
International School of Correspondence, Portland, Educational Literature.
Maccabees, Portland, Society Literature.
Montana State Commission, Montana, Educational Exhibit.
National Consumers' League, New York, Society Literature.
National Cash Register Co., Dayton, O., Social Economy Literature.
National W. C. T. U., Evanston, Ill., Literature.
Portland Women's Union, Portland, Society Literature.
Scientific American, San Francisco, Encyclopedias.
Stokes, Mary E., San Francisco, Oil Paintings.
Woman's College, Baltimore, Educational Literature.
Woman's Court, Minneapolis, Collective Women's Work.
Ward, M. J., San Francisco, Perfumery.

FLOOR

Australia Department, Collective Exhibit.
China, Collective Exhibit.
East India, per G. A. Hamilton, Collective Exhibit.
European (Foreign) Building.
Japan, per Y. M. Kushibiki, Collective Exhibit.
Turkey, Persia, Siam, Egypt, per Gaston Akoum, Collective Exhibit

EUROPEAN (FOREIGN) BUILDING

Austria, S. Herlinger, O. Moser.
France, Victor Laruelle.
Germany, Department.
Holland, Capt. M. J. Perk.
Hungaria, M. E. Fischer.
Italy, per P. Rossi, J. Zeggio.
Russia, M. Berkowitz.
Switzerland, Wm. Groth.

THE gates open to the public at 8 a. m. They remain open continually all morning, afternoon and evening.

Be sure to count your change when buying tickets to any attraction. While ticket-sellers are invariably honest, histories of expositions show that thousands of dollars are left behind by excited or absent-minded people.

Things Worth Knowing

Do not hesitate to ask information of any guard or employe of the exposition. They are paid for telling you things you wish to know concerning the exposition.

There are four entrances to the exposition. One is at each side. The main entrance is at the south. The north entrances give access from the river. Boats run from the various docks in the city to the exposition grounds. The trip by steam or gasoline boat requires from fifteen minutes to half an hour.

The exhibit palaces close at sundown. The time to see the exhibits is in the daytime. The Trail will be at its best by night. Nightly electrical illumination are a feature of the fair.

THE first time you see the exposition go in by way of the main entrance. You will thus be able to get a better idea of the plan on which the grounds are laid out. There is nothing complex about this plan.

Key to the Exposition

Passing through the main entrance, the first building to your left is the Administration building. This is merely the office-place of exposition officials.

Go straight ahead, passing under the Colonnades. That building to your right, this side of the Colonnades, is the Postoffice building, where the methods of handling United States mail are shown.

Free postal delivery is made herefrom to every part of the exposition.

Passing the Colonnades, the first building is the Fire Department, where model apparatus is installed and expert fire-fighters and trained horses stand ready to answer a summons of fire at an instant's notice. You are welcome to inspect this building.

The next building, with the latticed windows, is police and detective headquarters. This is the place to tell your troubles if you have any.

Having crossed the street that now intervenes, you are at the magnificent Agricultural palace. To the right is arranged the Machinery, Transportation and Electricity, the Liberal Arts, Varied Industries and Manufactures, and the Mining buildings, the Auditorium and the Missouri State building.

Grouped behind the Agricultural palace are the New York, Illinois, Idaho and Utah State buildings.

Returning again to the front of the Agricultural building, which is the logical starting point, the line of buildings to the left are the Foreign Exhibits, Oriental and Forestry buildings, the latter standing at the top of the gradual rise. South of these structures, arranged in a line, are the Coos County, Oregon, and Massachusetts buildings and the Museum of Fine Arts.

Adjoining the Forestry building are the California and the Washington State buildings, and west of these the beautiful natural park. Passing east again on one of the numerous paths, the main terraces appear. Here is given an exceptional view of the lake and lower part of the exposition.

Hereafter the situation is self-explanatory. That shell-shaped structure below you is the acoustical bandstand, where daily open-air band concerts will be given. To the right lies The Trail, and across the Bridges of Nations is the magnificent Government building.

Don't Fail to See

Hungarian Csarda

Little Hungary

The elite place of the grounds
Situated on the lake
Best view of lake and submarine light display
First-class European & American kitchen
Hungarian specialties
Best wines, beers & spirits guaranteed
Very reasonable prices

The Trail

THE Trail, the amusement feature of the exposition, marks the beginning of the bridge which crosses Guild's lake. After examining any of the thousands of interesting exhibits contained in the large palaces, the visitor may find much to amuse him within a few short steps of the principal parts of the exposition grounds. The Trail is easily reached from the Government buildings, from the American inn, Columbia court, or the Concourse plaza. Its attractions are all of the best, affording the cleanest and most spontaneous of humor. The management of the exposition has striven to make The Trail superior to the corresponding features of any previous exposition. Applications for concessions of an objectionable nature were denied, referring, in particular, to those of a morbid character. Like all features of the exposition, The Trail is compact. To see The Trail does not require a tiresome walk. Such was not the case at previous expositions. While most of the concessions may correctly be classified as amusements, a number of them are extremely edifying. Some offer magnificent exemplifications of plastic art, portraying the great calamity which overtook Galveston, the marvelous Cascades, of the St. Louis exposition, and a number of places of public interest. Others show the methods of mining in far northern Alaska, a country with which most of us are woefully unfamiliar. Animal phenomena are well represented and offer mystifying feats of calculation and acrobatics. Remarkable showings of the power of man over ferocious beasts of the jungle may be seen in the wild animal show.

The foolish side of The Trail makes a creditable showing with mazes of varied design, the Streets of Cairo, where Oriental maids in ethereal garb dance their peculiar terpischore to the wails of odd instruments; the Japanese Village, where the vistor may sip a cup of tea true to the customs of the Island Kingdom; Gay Paree, and various other places when fun is the chief feature.

UNDER the auspices of the Multnomah Amateur Athletic Club, and the supervision of H. W. Kerrigan, the athletic events at the Lewis and Clark Exposition will be held in Recreation park, only a short walk from the Palace of Machinery, Electricity and Transportation. The events will include inter-scholastic and inter-collegiate track and field meets and championship matches. Mr. Kerrigan has prepared an elaborate program for Indian day, which will offer all of the sports peculiar to our aborigines. Prominent athletes from all over the country will participate in the events, and the competition for awards will be keen. Six thousand dollars has been appropriated for medals.

Athletics

FESTIVAL HALL, popularly known as the Auditorium, is located on Concourse plaza, opposite the Palace of Manufactures, Liberal Arts and Varied Industries. In this will be held, during the exposition period, all public functions and the meetings of organized bodies, concerts and organ recitals. On Sunday afternoons the Auditorium will be the scene of religious services, some of the country's foremost ministers, of various beliefs, having volunteered to participate. The Mormon choir from Ogden will be heard here in a number of concerts. Also, under the auspices of the Nord-Pacific Saengerbund, a great saengerfest will be held. The Auditorium is substantially constructed. It seats 2500 people.

Auditorium

Principal Attractions and Prices of Admission

Ride in Launch		$0 25
Ride in Gondola		.50
Rowboat, one hour		40
Roller Chair, one hour		.60
Seating—Music Band Stand		10
Seating—Life Saving Exhibit		.10
Seating—Fire Works		10
Jabour's Gay Paree Vaudeville Theatre	On Trail	25
Concert Hall	On Trail	.25
Infant Incubators	On Trail	25
Cascade Gardens and Terrace of States	On Trail	.25
White Swan	On Trail	.10
Baltimore Fire	On Trail	.10
Water Chutes	On Trail	.10
Temple of Mirth	On Trail	.10
Japanese Village and Theatre	On Trail	25
Diving Elks and Princess Trixie	On Trail	.25
Jabour's New York Animal Show	On Trail	.25
Official Guide		.25
Haunted Swing	On Trail	.25
Cairo	On Trail	.15
Cairo Theatre	On Trail	.25
Cairo Ride on Donkey	On Trail	.25
Cairo Ride on Camel	On Trail	.30
Kiralfy's Carnival of Venice	On Trail	.50
Land of the Midnight Sun	On Trail	.25
Darkness and Dawn	On Trail	.25
Klondike Mining Exhibit	On Trail	.25
Televue	On Trail	.25
Davenport Farms Exhibit	On Trail	.25
Sistine Madonna	Foreign Building	.10
Trip to Niagara Falls	On Trail	.25
Official Daily Program		.05
Aerodrome	On Trail	.25
Aerodrome Ride in Balloon or Airship	On Trail	.75
Galveston Flood	On Trail	.25
California Radium Exhibit	On Trail	.25
Burns' Cottage	On Trail	.10
Haunted Castle	On Trail	.10
Trip to Venus	On Trail	10
Official Catalogue		.25
Shooting Gallery		.10
Mirror Maze	On Trail	.10
The White Slave, or a Glimpse of the Harem	On Trail	.10
		$10 00

FREE—Attractions You Should Not Miss

Free Stereopticon Lecture with Motion Pictures, hourly after 10 a m.	N C R. Building
Indians at Work Weaving Blankets	Mnfg. Building
Indians Making Baskets, Pottern and Jewelry	Mnfg. Building
Indian Blanket Valued at $5,000	Mnfg. Building
The Bubble Fountain	Mnfg. Building
The Electric Fountain	Mnfg. Building
The Making of Pocket Knives from the Raw Material	Mnfg Building
The Marvellous Automatons	Mnfg Building
The Making of Stiff Hats	Mnfg Building
Rope Works in Operation	Mnfg Building

MAP
SHOWING
CONCESSIONS
ON
TRAIL
TO AMERICAN INN
AERONAUTIC CONCOURSE
TO AMERICAN INN
ESPLANADE
HYDRAULIC RAM
KIRALFY
CARNIVAL OF VENICE
DAVENPORT FARM
EXHIBITS
ALASKA
&
KLONDIKE
GALVESTON
FLOOD
TEMPLE
OF
MIRTH
INFANT
INCUBATORS.
JAPANESE VILLAGE
THE TRAIL.
BRIDGE
CASCADE
GARDENS
Co.
STREETS OF CAIRO
ORIENTAL VILLAGE
NEW YORK
ANIMAL SHOW
ANIMALS
OTTO KUHLER
SIBERIA RESTAURANT
OLD PLANTATION
SHOOTING
GALLERY
WATER
CHUTES
WATER
PLATFORM
ESPLANADE

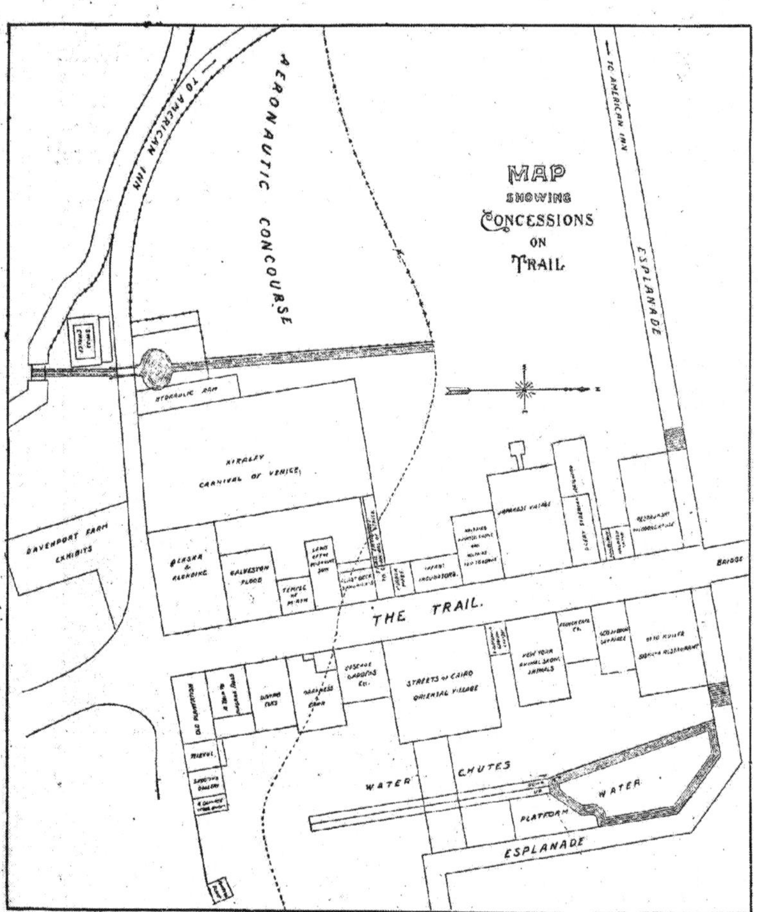

MAP OF TRAIL

DON'T fail to visit the Olympia Brewing Company's Swiss Chalet near St. Helen's Road entrance and get a drink of their famous mineral water, FREE FOR THE ASKING

¶ This water is from the Company's wells at the brewery in Tumwater, Washington, and is the same as used in making the famous OLYMPIA BEER. ¶ "It's the water" that gives Olympia Beer its character. ¶ Olympia Beer is to be had at first-class bars and restaurants in Portland. Ask for it and be convinced of its superiority. It is considered the best beer made on the Coast and is as good as can be brewed.

OLYMPIA BREWING COMPANY
OLYMPIA, WASHINGTON

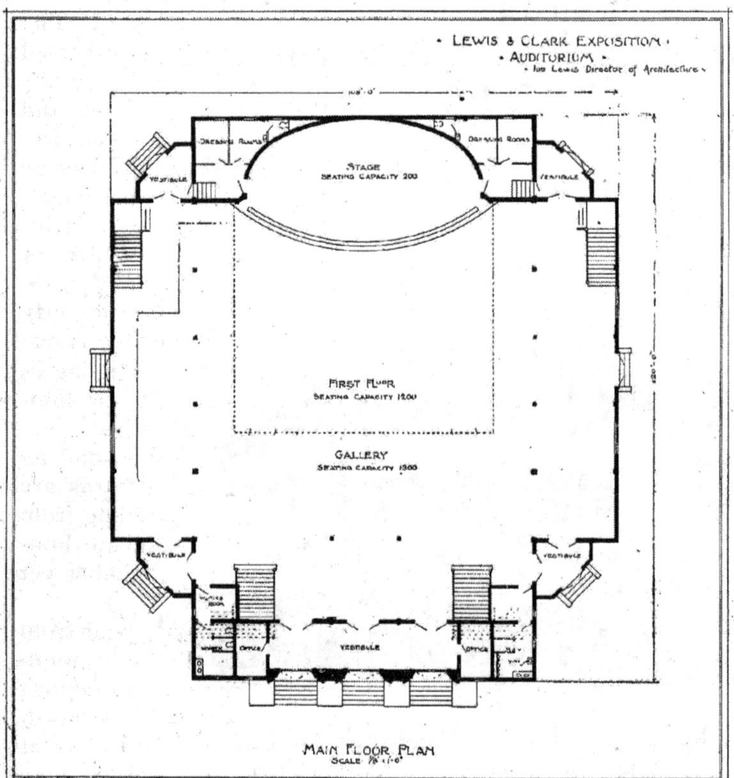

MAP OF AUDITORIUM

Lines of the Southern Pacific

VISITORS who have failed to see the country tapped by the Southern Pacific lines between Portland and San Francisco have missed the richest part of the great Western Empire. From Portland south as far as Ashland is one vast field of opportunities for the homeseeker and investor. The famous Willamette valley is traversed its entire length in making this trip. The hop-growing sections of Marion and Linn counties, the grain and fruit-producing sections of Lane county may be passed through in daylight. In both counties the industries are not developed to the extent they will be in future years, and the enterprising man may be a part of the future development and profit by it.

From the Willamette valley the road passes directly into the center of the Umpqua valley. This valley is one of the richest in the West. The mountains surrounding it, while they have yielded up thousands of dollars in mineral wealth, are yet but partially developed.

The Rogue River valley is next encountered, and its vast fields of green alfalfa and flourishing orchards are traversed from end to end. A new railroad, building from Medford to the country farther east, will open up thousands of acres of virgin soil and vast areas of timber yet unscarred by the woodman's axe.

Ashland is in the very head of the valley, and from here commences a laborious climb over the high mountain range. The scenery of the Shastas is famous throughout the world and needs no description. Descending the mountain, the train stops for several minutes at the Shasta Mineral Springs, where the traveler may alight and drink nature's natural soda as it sparkles from the ground. There are two trains daily over this route, so that the entire trip may be made in daylight.

FLOOR PLAN OF FORESTRY BUILDING

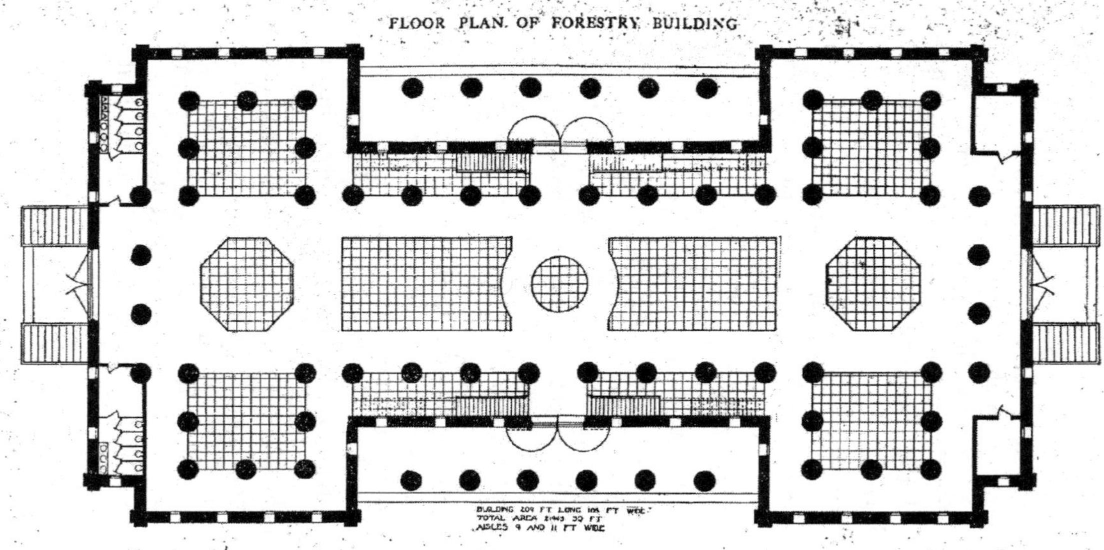

IN FOREIGN EXHIBITS BUILDING ADJOINING BRITISH SECTION NORTHWEST CORNER OF ROOM

This wonderful bit of needle painting is probably the greatest achievement of woman from an artistic view, and has challenged the admiration of the whole world. It received the Gold Medal at the Paris Exposition in 1900, and the highest award—the Grand Prize—at the Louisiana Purchase Exposition, at St. Louis, in 1904. Its art value is estimated at $50,000.

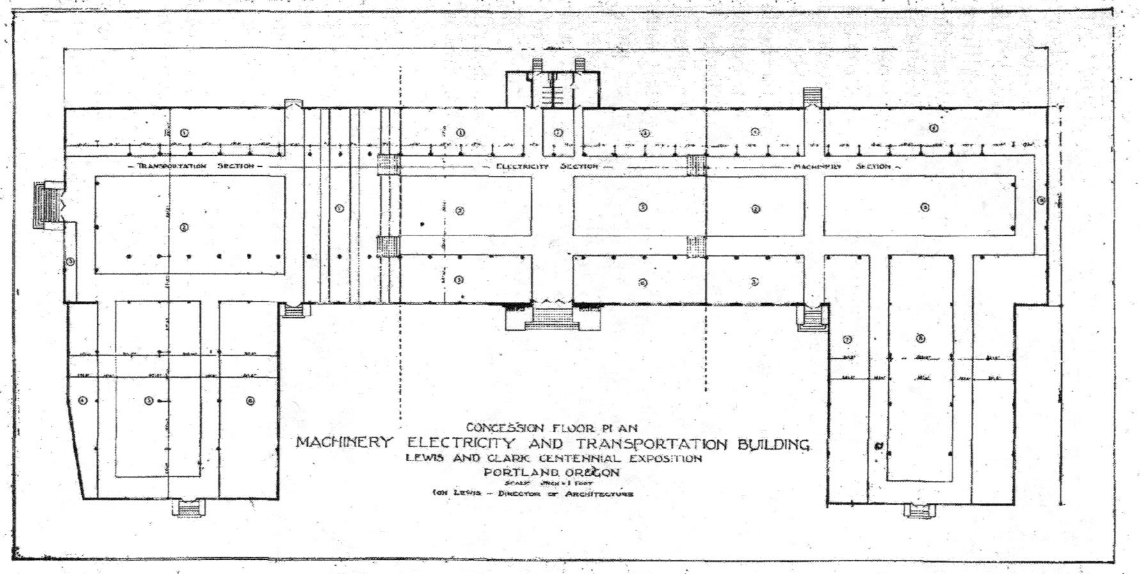
TRANSPORTATION SECTION
ELECTRICITY SECTION
MACHINERY SECTION
CONCESSION FLOOR PLAN
MACHINERY ELECTRICITY AND TRANSPORTATION BUILDING
LEWIS AND CLARK CENTENNIAL EXPOSITION
PORTLAND OREGON
ION LEWIS – DIRECTOR OF ARCHITECTURE

The United States Government Exhibit

NOT to see the Government's buildings and exhibits is to miss a liberal education. All departments of the government are fittingly represented by elaborate displays that tell their own story.

Inside the main building you should see: Heavy artillery, in charge of artillerymen and navy gunners; water buffalo and other Philippine exhibits; navy hospital and collection of projectiles used by the big guns; army pack-train, hospital wagons, arms and ammunition, including a collection that shows the evolution of firearms from the old matchlock to the latest magazine rifle; model of famous Arlington cemetery; models of postal cars, famous warships, canal-locks, fortifications and the section showing methods of postal delivery from Malamoot sled team to latest mail car.

Decorations on the walls show Yellowstone park and government buildings and utilities and paintings of famous statesmen. In an ante-room are stuffed animals from all over the world, together with remains of prehistoric animals. The complete skeleton is here shown of the largest mastadon ever unearthed in America. There is also a meteor that fell in Mexico a year ago and which weighs four tons. Best methods in agriculture are fully shown. Attaches of the government attend each exhibit and will devote their time to making any and all explanations that may be sought by visitors regarding any phase of the exhibits. Important government works that cannot be shown in exhibits are presented in life-motion pictures. Naval and land battles will be presented in this manner.

Outside the main building are totem poles and many picturesque and interesting exhibits too large for the interior. A full display from Alaska will be found in the Alaskan wing. Another interesting display is shown in the Fisheries wing. A fully equipped life-saving station stands on the banks of the lake near the main building.

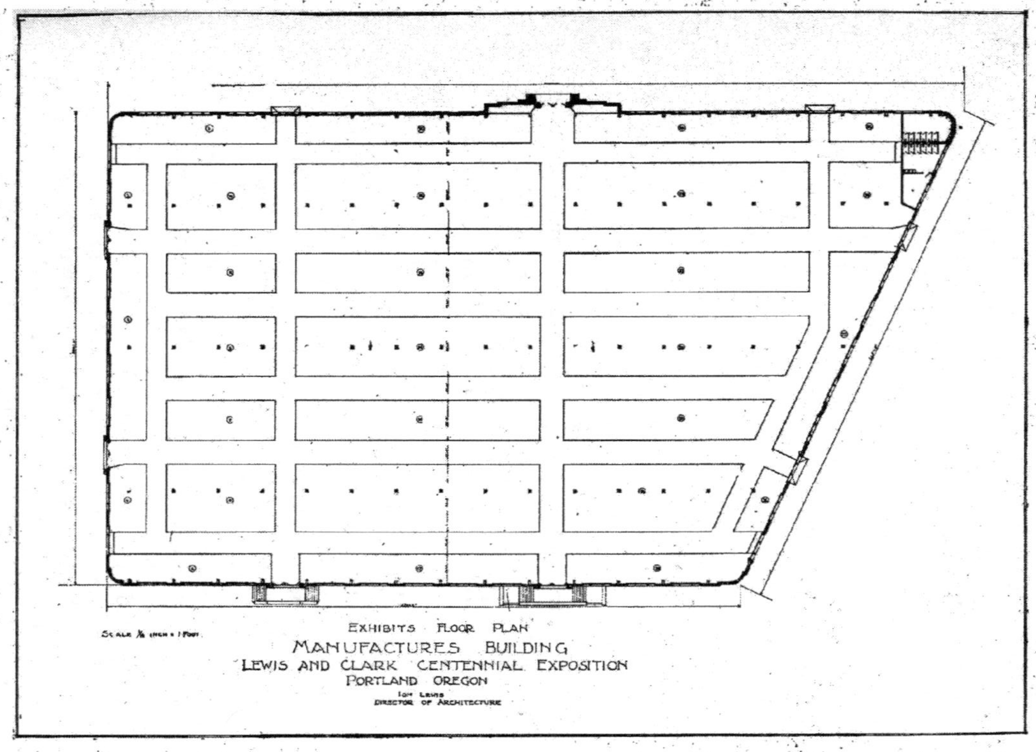
EXHIBITS FLOOR PLAN
MANUFACTURES BUILDING
LEWIS AND CLARK CENTENNIAL EXPOSITION
PORTLAND OREGON
ION LEWIS
DIRECTOR OF ARCHITECTURE

Spokane, Washington

Grown to 75,000 population in twenty years
Greatest railroad center west of Chicago
Seven wonderful water falls—150 feet
Biggest lead mines in the United States
160 miles of electric lines
A city of schools and homes
Fine business blocks
Comfortable Hotels
Magnificent Restaurants
Beautiful lakes and mountain scenery
Golf, tennis and bathing

Your ticket permits stopover

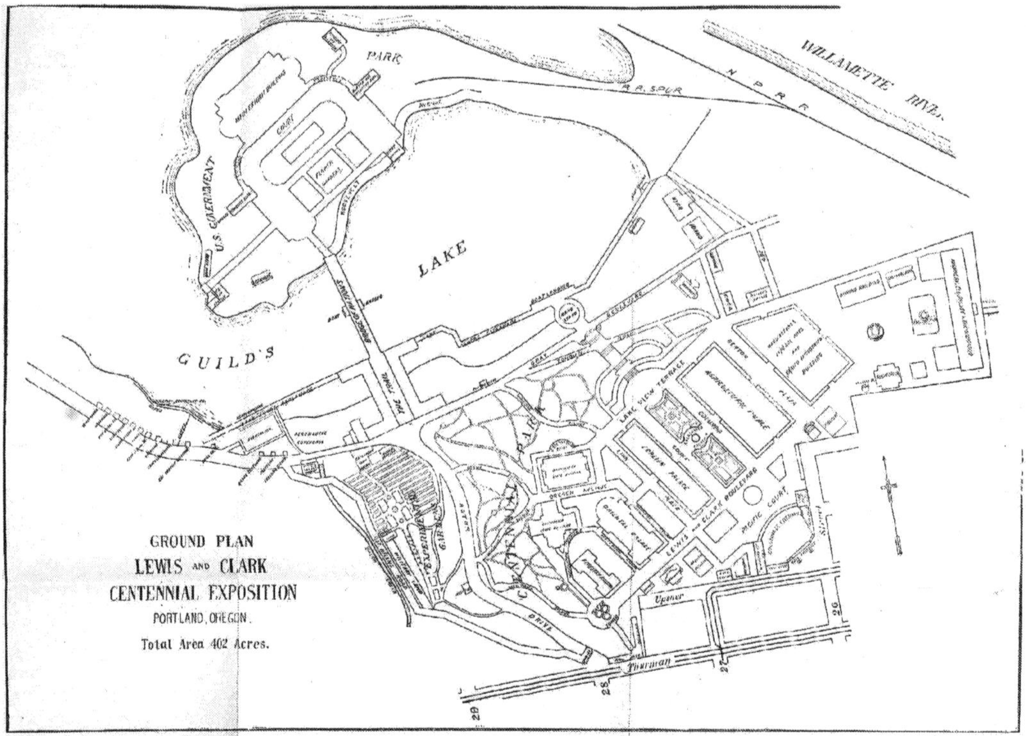
PARK
WILLAMETTE RIVE
R.R. SPUR
N P R R
U.S. GOVERNMENT
LAKE
GUILD'S
LAKE VIEW TERRACE
LEWIS AND CLARK BOULEVARD
PACIFIC COURT
Upshur
Thurman
GROUND PLAN
LEWIS AND CLARK
CENTENNIAL EXPOSITION
PORTLAND, OREGON.
Total Area 402 Acres.

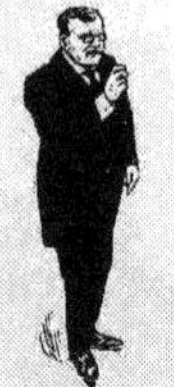

Roosevelt Said:

"I never saw two such cities anywhere as Spokane and Seattle. If my eldest boy was large enough to be choosing a place, I would advise him to locate in one or the other of those cities and it is a shake-up between them."

—President Roosevelt.

ONE OF THE SEVEN FALLS AT SPOKANE, WASH.

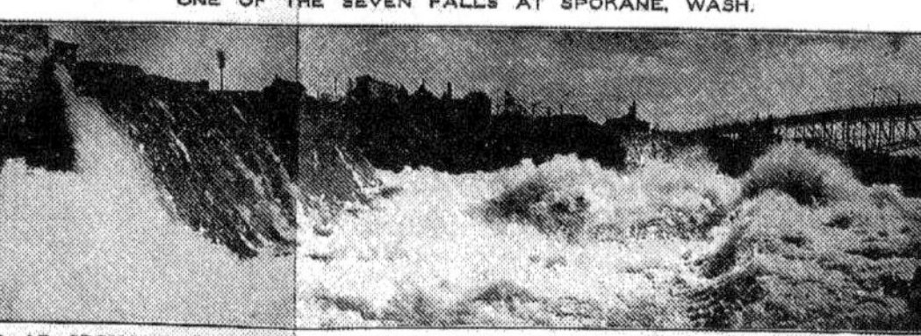

YOUR TICKET ENTITLES YOU TO STOP OVER AT SPOKANE, THE MOST BEAUTIFUL CITY OF THE WEST

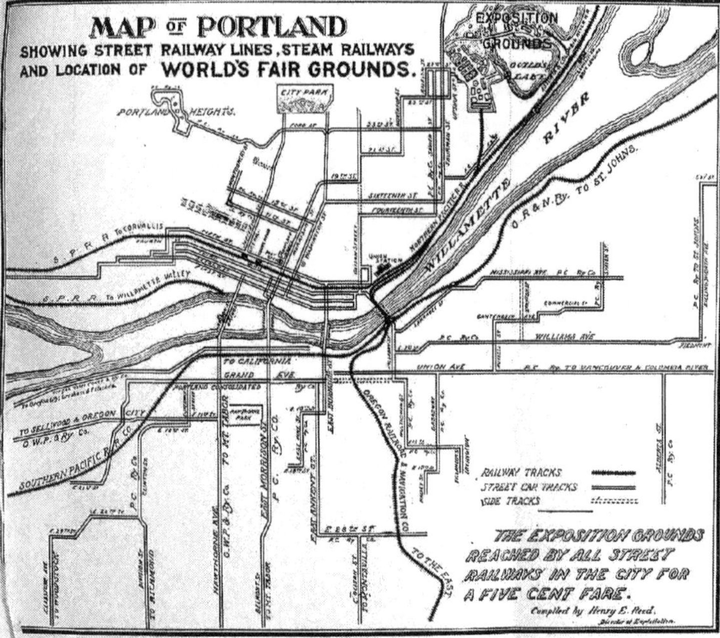

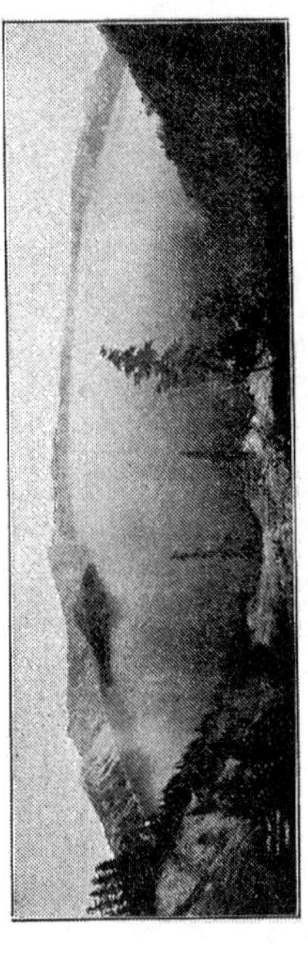

CRATER LAKE
ONE OF THE WORLD'S NATURAL WONDERS -
REACHED FROM MEDFORD, OREGON

www.ingramcontent.com/pod-product-compliance
Lightning Source LLC
LaVergne TN
LVHW011125210726
843642LV00016B/480
9781018742465